WITHDRAWN FROM
Greenville (SC) Co. Library System

# LET'S ROCK!

EXPLORING WEATHERING AND EROSION

MARIE ROGERS

PowerKiDS press

New York

Published in 2022 by The Rosen Publishing Group, Inc.
29 East 21st Street, New York, NY 10010

Copyright © 2022 by The Rosen Publishing Group, Inc.

All rights reserved. No part of this book may be reproduced in any form without permission in writing from the publisher, except by a reviewer.

First Edition

Portions of this work were originally authored by Maria Nelson and published as *Weathering and Erosion*. All new material in this edition authored by Marie Rogers.

Editor: Amanda Vink
Cover Designer: Alan Sliwinski
Interior Designer: Rachel Rising

Photo Credits: Cover kavram/Shutterstock.com; Cover, pp. 1, 3, 4, 5, 6, 7, 8, 10, 12, 13, 14, 15, 16, 18, 20, 21, 22, 23, 24 (background) Alex Konon/Shutterstock.com; p. 5 VectorMine/Shutterstock.com; p. 6 Jumnong/Shutterstock.com; p. 7 Nicole Glass Photography/Shutterstock.com; p. 8 Feifei Cui-Paoluzzo/ Moment/Getty Images; p. 9 Westend61/China/Getty Images; p. 10 Timothy Lee Lantgen/Shutterstock.com; p. 11 Dusan Milenkovic/Shutterstock.com; p. 12 Michael Zysman/Shutterstock.com; p. 13 Suttipong Sutiratanachai/Moment/Getty Images; p. 15 Eloi_Omella/E+/Getty Images; p. 16 Piyawat Nandeenopparit/Shutterstock.com; p. 17 PhotoQuest / Contributor/Getty Images; p. 18 zhang kan/Shutterstock.com; p. 19 Geography Photos / Contributor/Getty Images; p. 20 SChompoongam/Shutterstock.com; p. 21 Richard Whitcombe/Shutterstock.com; p. 22 Daniel Cicivizzo/EyeEm/Getty Images.

Some of the images in this book illustrate individuals who are models. The depictions do not imply actual situations or events.

Cataloging-in-Publication Data

Names: Rogers, Marie.
Title: Exploring weathering and erosion / Marie Rogers.
Description: New York : PowerKids Press, 2022. | Series: Let's rock! | Includes glossary and index.
Identifiers: ISBN 9781725319332 (pbk.) | ISBN 9781725319356 (library bound) | ISBN 9781725319349 (6 pack)
Subjects: LCSH: Weathering–Juvenile literature. | Erosion–Juvenile literature. | Geochemical cycles–Juvenile literature.
Classification: LCC QE570.R5936 2022 | DDC 551.302–dc23

Manufactured in the United States of America

CPSIA Compliance Information: Batch #CWPK22. For further information contact Rosen Publishing, New York, New York at 1-800-237-9932.

CONTENTS

EARTH'S LANDSCAPES 4

WEATHERING 6

EROSION. 10

IN THE ROCK CYCLE. 14

DANGER! 16

SPEEDING IT UP 18

SOIL EROSION 20

CHANGING THE WORLD 22

GLOSSARY 23

INDEX 24

WEBSITES 24

EARTH'S LANDSCAPES

From the bottom of the ocean to the top of Mount Everest, rocks are all around us! Rocks are one of the building blocks of Earth. They're solids made of one or more minerals, or naturally occurring nonorganic materials. Rocks are always slowly changing. They break down and reform as new rocks. That's because of processes such as weathering and erosion.

Weathering and erosion work hand in hand. These two processes turn large mountains and **boulders** into sediment, or tiny pieces of rock. Then sediment becomes something new in the rock cycle. Weathering and erosion are two of the reasons Earth looks the way it does!

Earth's many landscapes are created by the rock cycle, including weathering and erosion.

ROCKING OUT

The rock cycle is a **concept** that helps us understand how rocks change because of their surroundings. There are three main types of rocks: sedimentary, metamorphic, and igneous.

Weathering and Erosion

Runoff

IGNEOUS ROCK

SEDIMENTARY ROCK

METAMORPHIC ROCK

WEATHERING

Weathering and erosion often work together, but it's important to remember they're two different processes. Weathering is the breakdown of rock into sediment. Sediment may then mix with plants and animal remains, **fungi**, and bacteria. This mixture becomes soil. Fertile soil, which has a mixture of many minerals, is soil in which plants grow well.

Physical weathering, or mechanical weathering, changes a rock's size or shape by force. This can happen when water flows into open spaces in rock. Physical weathering can also happen when wind blows matter at rock and chips little bits of it away.

ROCKING OUT

Changes in **temperatures** can weather a rock. When water freezes into ice, it gets bigger and takes up more room. The ice works as a wedge to widen cracks in rock.

Animals can also cause physical weathering. While it's digging, this prairie dog may chip away rock.

Adding new minerals, chemicals, or gases to a rock may change its makeup and cause it to break down. This is known as chemical weathering. Rain is a common way chemical weathering occurs. Rainwater mixes with chemicals and minerals as it falls. Then it flows over and into rocks.

Biological weathering happens because of living things. For example, a tree's growing roots may weaken the rocks around it. There are also some organisms, such as bacteria, that break down rock. Lichen, a mix of fungi and **algae**, makes rock its home and slowly wears it away over time.

LICHEN

The Shilin Stone Forest is an area of limestone formations located near Kunming, China. The limestone has been dissolved, or melted away, by chemical weathering.

EROSION

Weathering creates sediment. Erosion can also create sediment while it's happening, but it's erosion that carries sediment away and **deposits** it in a new location. There are many natural **agents** of erosion.

Gravity, or the force that pulls everything toward Earth's center, is one major agent of erosion. Its force can pull pebbles, rocks, and boulders down cliffs and hills. This is called mass movement. Landslides and mudflows are examples. They both happen suddenly. Mudflows, also called mudslides, occur when there is a lot of water, perhaps after a period of heavy rain or after a lot of snow melts.

Mudflows can move as fast as 50 miles (80.5 km) an hour! They can cause a lot of damage, or harm.

Water is a major agent of erosion. Water both breaks down rock into sediment and carries it away. For example, ocean waves beat against the shoreline and pull bits of rock away. Rivers wear away their banks over time. For this reason, lakes, oceans, and other bodies of water have lots of sediment in them.

Wind is another agent of erosion. It blows around loose sand, soil, and other sediment. This can often create even more sediment! Mountains and other landforms might be "sandblasted," or weathered by sand that's being carried by wind.

ROCKING OUT

Climate impacts erosion too. In drier climates like deserts, chemical breakdown of rocks happens slowly. However, mechanical breakdown happens fast. That's because there are fewer plants, which help stop erosion!

Water and wind are agents of both weathering and erosion.

IN THE ROCK CYCLE

Weathering and erosion are a big part of the rock cycle. That's because sediment is needed to make sedimentary rock. Sedimentary rock is created when sediment piles up and, over time, **solidifies** into rock. Depending on what happens, sedimentary rock can become metamorphic rock or igneous rock when it's exposed to forces such as heat and pressure. All rocks can move anywhere in the rock cycle.

As steps in the rock cycle, weathering and erosion are part of the creation of many rocks on Earth. When rocks deep underground come up to the surface, they'll be weathered and eroded once again!

These rock formations in Zion National Park in Utah are made of sedimentary rock.

ROCKING OUT

Some rocks are deep underground, so deep they aren't exposed to weathering and erosion. Uplift pushes rocks to the surface. **Uplift** can happen during an earthquake.

DANGER!

Weathering and erosion can also be risky and even sometimes harmful for humans. Erosion can hurt crops. In very dry years, wind can blow away topsoil, or the fertile upper layer of soil. This happened on a large scale in the United States during the Dust Bowl in the 1930s. Farmers had planted wheat instead of the grasses that once grew there and left other fields empty. There was nothing to keep topsoil in place.

Heavy rains can also wash away soil. If the ground becomes saturated with, or full of, water, the soil can hold no more. **Runoff** flows to low areas and takes topsoil with it.

This picture from May 1936 shows a giant dust cloud created from soil erosion during the Dust Bowl.

SPEEDING IT UP

Weathering and erosion happen naturally, but human activities can speed them up. Fossil fuels, or fuels formed in the earth from dead plants and animals, are nonrenewable **resources**. That means once they're used up, they're gone. Humans burn fossil fuels to create energy. However, this can put chemicals into the air. Some of these chemicals turn acidic when expoed to sunlight. This means that the chemicals can **disolve** other objects. They fall back to Earth as part of acid rain.

Acid rain causes weathering to happen faster. It can wear stone away and harm lakes and rivers.

Acid rain can hurt old buildings and new plants.

SOIL EROSION

Humans regularly clear large areas of forest to make room for buildings, industry, and farmland. But deforestation, or the removal of large areas of trees and plants, can cause soil erosion too. That's because the roots of trees and other plants hold soil together. These roots help keep the levels of water within an area in balance too.

When soil erosion happens, too much sediment can get into waterways. This can harm waterways and affect our drinking water. Soil erosion can also increase the amount of dust carried by wind, which can spread diseases.

ROCKING OUT

Human activities are major contributors to deforestation. Almost half of the world's topsoil has been lost to erosion. Humans need to find other, better ways of building.

Every second, an area of forest the size of a soccer field is lost to deforestation.

CHANGING THE WORLD

Weathering and erosion continue to change our landscapes. For example, the Grand Canyon was made through the processes of weathering and erosion. The Colorado River carved its way through sedimentary rock. This is called downcutting, and it results in the formation of canyons and valleys. This process continues to make the Grand Canyon deeper and wider. Someday, a long time from now, the Grand Canyon won't look like it does today.

Weathering and erosion will continue to break up rock. Next time you find a small stone or walk through sand, remember it was once part of something much bigger!

GLOSSARY

agent: Something or someone that causes something to happen.

algae: Simple plants with no leaves or stem that grow near water.

boulder: A very large stone or rounded piece of rock.

concept: An idea of what something is or how it works.

deposit: To put something down and leave it behind.

dissolve: To break down a solid when a liquid mixes with that solid.

fungus: A living thing that is like a plant but that doesn't have leaves, flowers, or green color or make its own food. The plural form is fungi.

resource: Something that can be used.

runoff: Water from rain or snow that doesn't go into the ground.

solidify: To make something solid or hard.

temperature: How hot or cold something is.

uplift: The upward movement of rocks to Earth's surface.

INDEX

A
acid rain, 18, 19
algae, 8
animals, 6, 7

B
bacteria, 6, 8
biological weathering, 8

C
chemicals, 8, 18
chemical weathering, 8, 9
climate, 13
crops, 16

D
deforestation, 20, 21
downcutting, 22
Dust Bowl, 16, 17

F
fossil fuels, 18
fungi, 6, 8

G
gases, 8
Grand Canyon, 22
gravity, 10

I
ice, 7
igneous rock, 5, 14

L
landslides, 10
lichen, 8

M
mass movement, 10
mechanical weathering, 6
metamorphic rock, 5, 14
minerals, 4, 6, 8
mudflows, 10, 11

P
physical weathering, 6, 7

R
rain, 8, 10, 16
rock cycle, 4, 5, 14
runoff, 16

S
sediment, 4, 6, 10, 12, 14, 20
sedimentary rock, 5, 14, 22
soil, 6, 12, 16, 17, 20

T
temperature, 7
topsoil, 16, 21

U
uplift, 15

W
water, 6, 7, 8, 10, 12, 13, 16, 20
wind, 6, 12, 13, 16, 20

WEBSITES

Due to the changing nature of Internet links, PowerKids Press has developed an online list of websites related to the subject of this book. This site is updated regularly. Please use this link to access the list:
www.powerkidslinks.com/letsrock/weatheringanderosion